Mixed Blessings

Rural Life on the Quantock Hills

G J Penny

In association with

First published in Great Britain 2007

Copyright © 2007 text and photographs Gary Penny

All rights reserved. No part of this publication may be reproduced, stored in a retrieval system, or transmitted in any form or by any means without the prior permission of the copyright holder.

ISBN: 978 - 0 - 9557802 - 0 - 2

Produced in association with The Quantock Hills Area of Outstanding Natural Beauty, with project funding from the Quantock Hills Sustainable Development Fund.

Published by:

Quantock Hills A.O.N.B.

www.quantockhills.com

G. J. Penny MA

www.wessexfoto.com

Printed and bound by B.A.S. Printers

Special thanks must go to the people of the farming and wider rural community on and around the Quantock Hills who have allowed me such generous access to their lives over the last couple of years, both to make pictures of them, but also to give up their valuable time to record many hours of interviews, so that future generations will be able to better know and understand what it was like to live and work around the Quantock's at the beginning of the twenty-first century.

This book and the project from which it is taken was made possible by the Quantock Hills Sustainable Development Fund, and the unflinching support of Ian Porter, Projects Officer, and all the team at the Quantock Hills Area of Outstanding Natural Beauty.

Introduction

The life of the farmer has always been a hard one, for those whose well-being is linked to the vagaries of this island's climate life is never going to be one of ease and relaxation. As hard as the life of the farmer is physically, for millennium's past it has at least been possible for a strong man or woman, or more usually both, to work hard and create a sustainable life for themselves and their family. Over the last hundred years or so we have seen mechanisation (and in our wider society industrialisation) threaten to make this view of the small, independent, family run farm a thing of the past. The reasons for this shift are huge and complex and not within my remit to explore here, however we can say that since the end of the Second World War, the number of small, family run, mixed farms has shrunk enormously.

In 1939 there were approximately half a million farms in Britain. The majority of these farms were of less than fifty acres, and supported around one and a half million families. There are now well under 200,000 farms of which about ten percent control more than fifty percent of the national output.

In our flatter counties we see great prairies stretching for miles and miles, without a hedgerow in sight anywhere. Where once there would have been hundreds of farms supporting many families, who in turn supported and were part of vibrant, living communities, containing a wide range of shops as well as schools and thriving small businesses, we now have a new breed of country dweller, the commuter, who likes to live in the countryside, so long as it doesn't smell badly or make too much noise first thing in the morning.

In the rural west of England we have been luckier than in many areas, by hanging on to a small farming way of life for longer than the average, but even here the pressures of modernity are changing the rural picture. The role of the farmer is now as much about preserving the natural environment for visitors as it is about farming. The economics of farming now make the idea of setting up as a small scale mixed farmer, and being able to support a family quite unthinkable. The old joke that the best way to make a small fortune in farming is by starting with a large one, and then buying a farm, is especially true of the small mixed farm today.

What we have seen is an increase in the number of hobby farms, bought by wealthy urbanites, who are able to combine a love of the countryside with a substantial second income or a large amount of disposable capital. Then there are the media types who buy an old farmhouse together with a few acres, a bit of woodland and some pigs, and then bombard us with articles in the broadsheets and with books about their 'farming' life. Are these 'hobbyists' helping the cause of the small farmer, or merely helping to disguise the problem?

Do we, as a society, still value those attributes that are required to be successful in farming, hard work, self-sacrifice a willingness to do without those material possessions that most of us today consider so essential. Since the Second World War, our society's obsession with the production of 'cheap' rather than good healthy food has made it harder and harder for the small farmer to survive against the agri-business giants, and the bulk buying power of the supermarkets.

Since 1945, the true cost of this 'cheap' food has been the massive reduction in the number of working mixed farms and the wider social networks that they helped to support. Unless we start to appreciate the enormous social advantages that come from maintaining and building thriving communities of families working and living in the English countryside, and unless this happens soon, the number of those families who are able to continue the fight will keep shrinking, and our rural environment will simply become a managed theme park for the recreation of our ever growing urban population. If we continue to let this happen, one day soon, we may well look back and see that we have lost some of the finest people our little island has bred, and we will all, as a nation, be the less because of it.

70% of the UK is Agricultural land

Janet White, sheep farmer and author, and her son Robert

Janet: I came to Durborough farm in 1966, with four small children, before that I had been a shepherd and share farmer in New Zealand, I have always loved sheep and hill country and from an early age worked in farming. When I was first married we had run a smallholding where the children were born, when I came here in 1966, Robert, who is now my farming partner was a six month old baby. I sometimes wonder whether you like sheep as much as I do?

Rob: Some days…

Janet: We also have a herd of beef cattle, it's an all grass farm of 250 acres with grazing rights on the common, Robert gets quite involved with the pony breeders and the annual rounding up, which is quite a good social event and good fun.

I'm 76 now so I find the hardest part is the shearing, I taught all my children to shear, I did about fifty this year, but Robert is now about twice the speed that I am. I can remember as a girl shearing Scottish black face sheep by hand, I would do fifty in a day, which included catching the sheep, rolling the wool, and that included going out at four o'clock in the morning to bring the sheep down off the mountain, but I can't do as many as that now.

When I farmed in New Zealand fifty years ago my annual wool cheque was £400 which was the wool from two hundred ewes, and that was enough for me to live on for a year, and buy another hundred sheep! We now clip twice as many sheep, and we receive about £500 from those fleeces, which is enough to support us for about ten days!

When we first came here there were five farms in the valley, and the people from all those farms would get together at Aisholt church. Now out of those five farms Durborough is the only one that is left as a working farm, all the other farm houses have been split off from their land and are now just houses, I think that's very sad. The farmers were very much the backbone of the community, my late husband was a churchwarden and chairman of Somerset NFU. I think people feel a bit more distant now and of course we are all kept so busy.

Rob: I don't have time for much, and I think there's a lack of other local farmers to talk to. Once upon a time you could chat to another farmer over the hedge, that's now gone, now we are all using mobile phones or sending each other text messages. Despite all the negatives, we still take time when we can, to stop on the hill on the quad bike and just take in the beauty of this place, and to appreciate all the wonders of nature. Whether it's a kingfisher by the pond, the deer on the hill or even a snake in the yard! I know a lot of young people would like to go into farming, but despite the beautiful scenery you can't live on fresh air.

Janet: It's a wonderful spot, it's got one sunny side, one damp side, north facing and a south facing valley and a stream running down the middle and heather on the top of the hill. When I said as a young girl that I wanted to be a hill farmer, they said, there's no money in that, and I said well I want to be a writer too, and they said there's no money in that either, and now I've done both, and they were right, but I wouldn't have changed my life for anyone's.

Rob and Janet White at Durborough farm, Aisholt.

Janet White feeding ewes at Durborough farm, 2006.

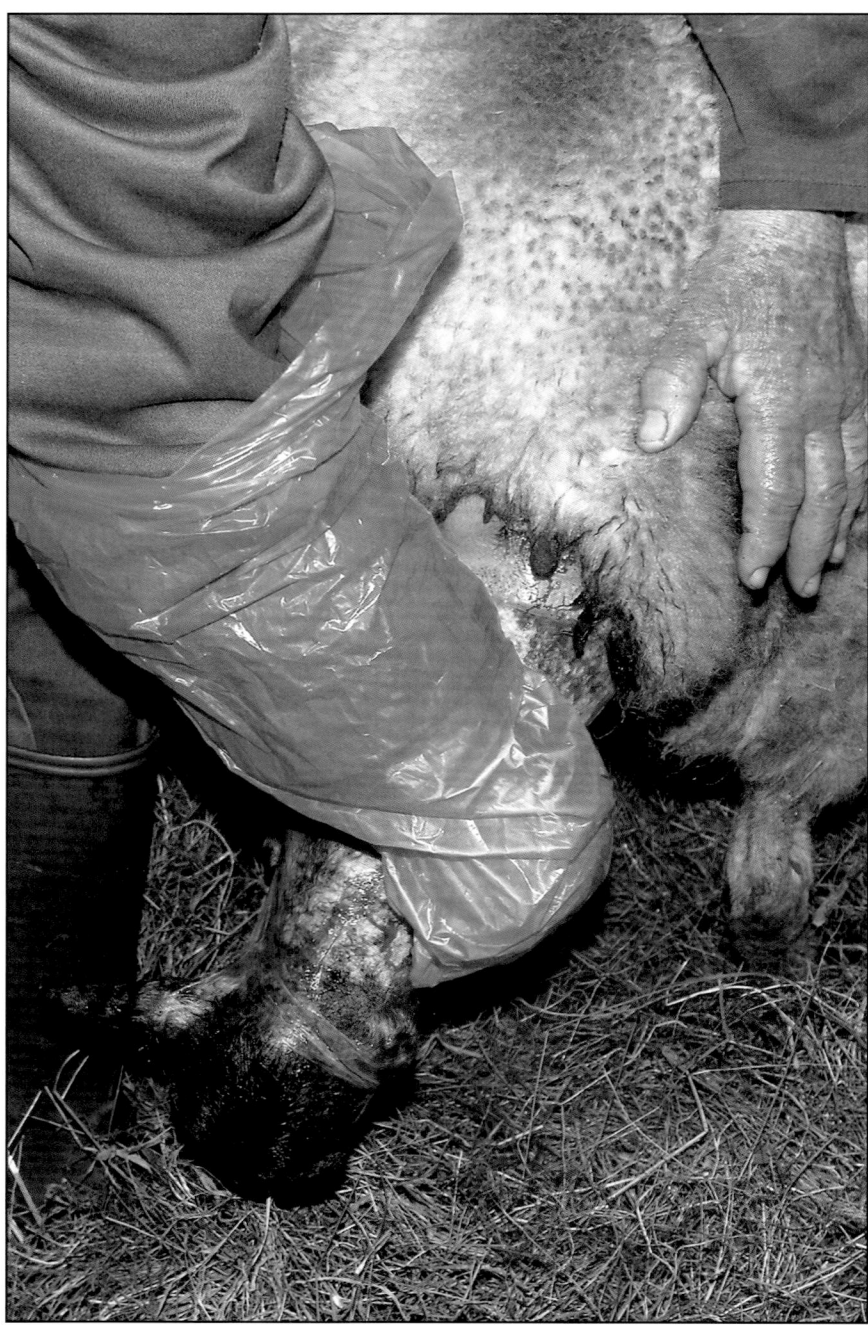
Janet White helping a new lamb into the world.

Preparing bottle formula for orphaned lambs in her kitchen at Durborough farm.

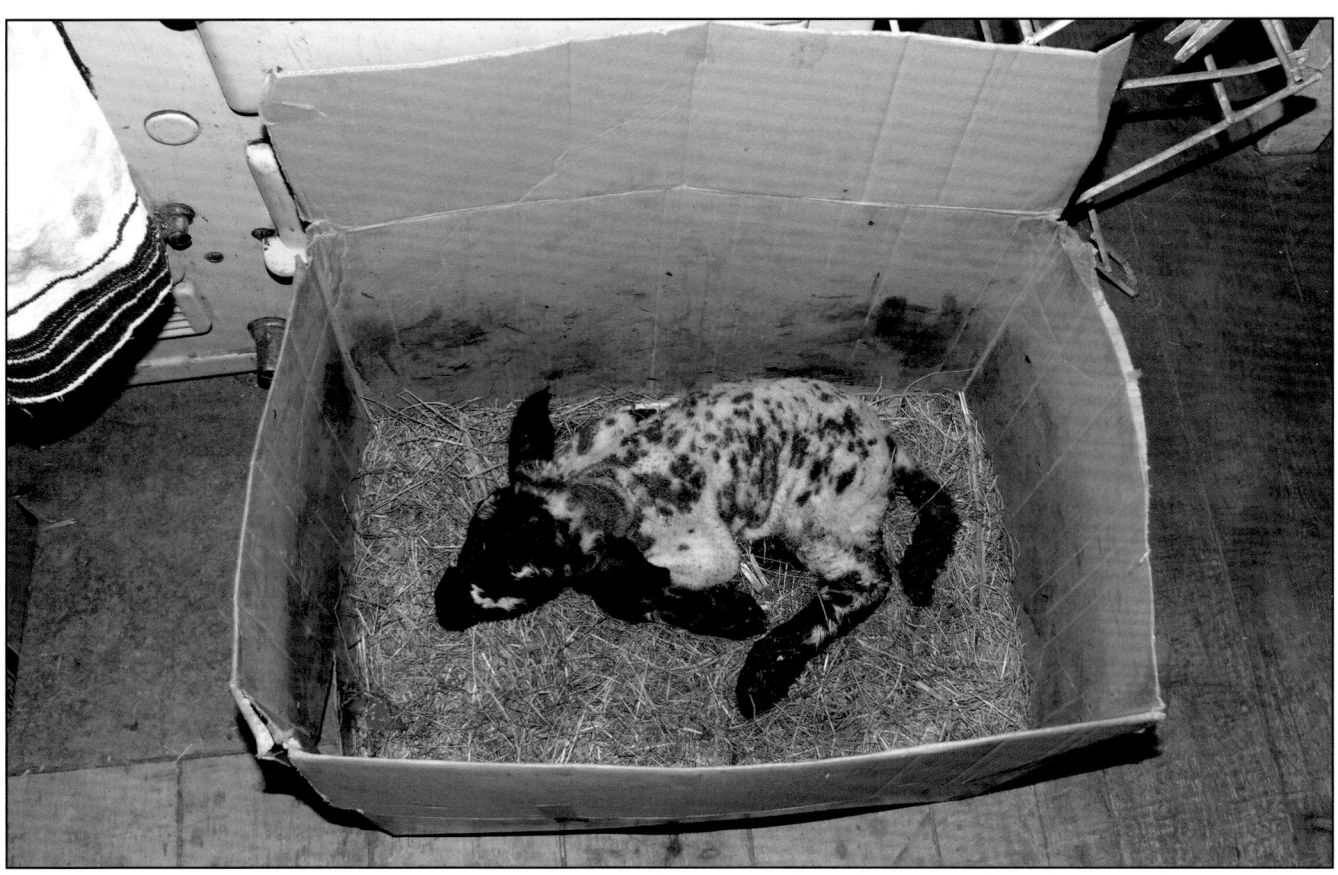

One day old orphaned lamb keeping warm by the aga at Durborough farm.

Janet White rounding up sheep for shearing, Durborough farm, summer 2006.

Mother and new lamb at Durborough farm, 2006.

Janet and Rob White shearing ewes, summer 2006.

Black faced ewe at Durborough farm, 2006.

Ingerma Rossiter at Waterhouse farm, summer 2006.

Jim and Jack Rossiter at Waterhouse farm, summer 2006.

Jack Rossiter on his sixty five year old harvester/binder cutting wheat for reed thatching at Waterhouse farm, summer 2006.

Ingerma Rossiter stacking freshly cut 'stooks' of wheat to dry in the sunshine, summer 2006.

Mark, Peter and David House (back row), Ella Sheldon and her daughter Jane House with Jane's new daughter-in-law Lorraine House, outside the newly renovated Tetton farm, summer 2006.

Jane House watching silage making at Tarr farm, summer 2006.

Making 'small' hay bales at Tarr farm, summer 2006. Although much slower to produce than the more modern big bales, the smaller ones are more portable, and find a ready market with horse owners.

Mark House helps guide his father Peter, as he moves their brand new combine through the barn up to one of their fields to be used for the first time.

Peter House driving the combine on its first day of cutting barley at Tarr farm, summer 2006.

The yard at Staples farm, West Quantoxhead, March 2006.

Nick Gibbons of Staples farm, West Quantoxhead, 2006.

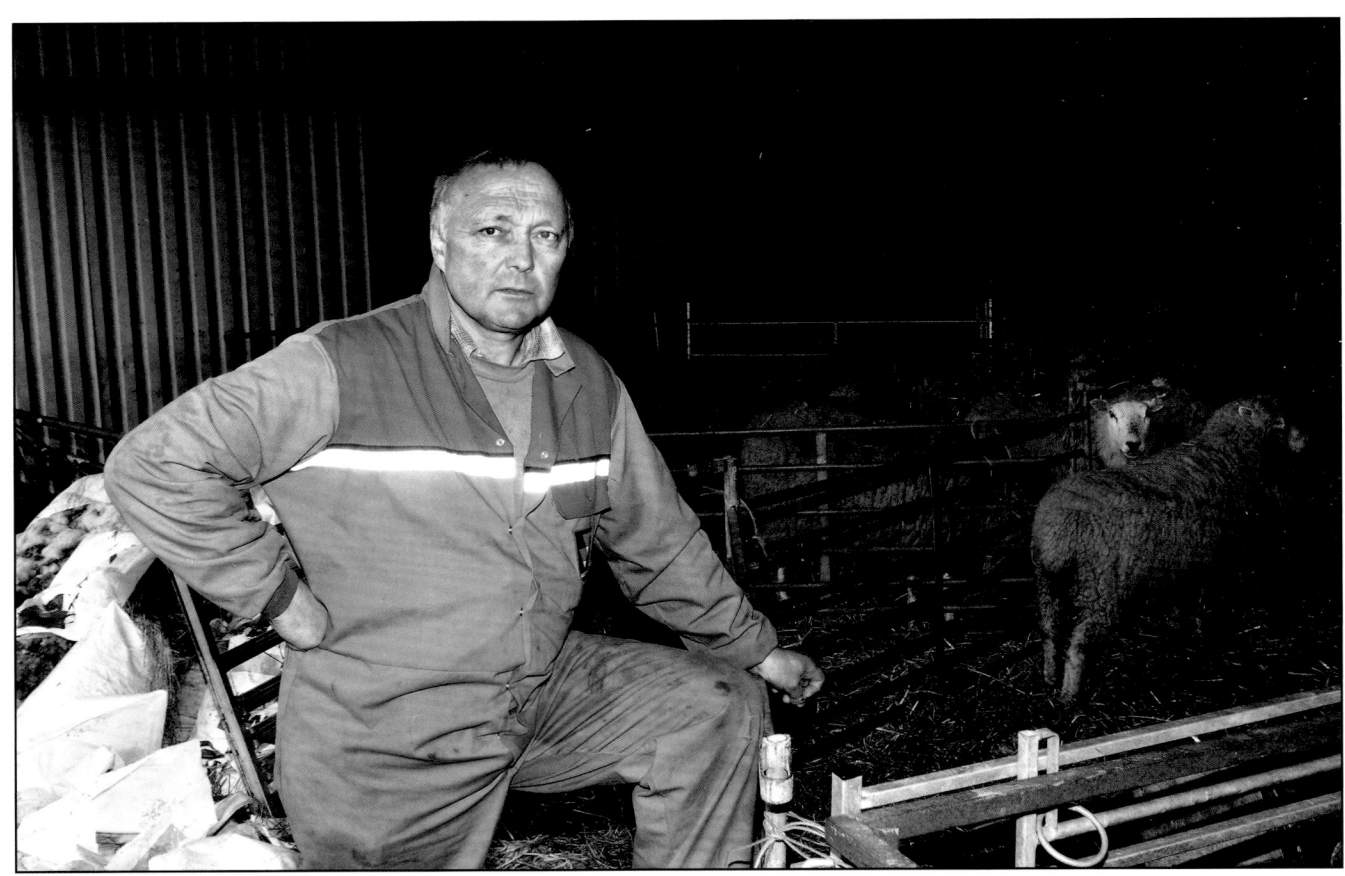

Nick Gibbons, Staples farm.

Chris Gibbons of Staples farm. A keen member of the Quantock Young Farmers Association, who sees hope for the future of mixed farming.

Graham Bigwood, ex-dairy farmer, who once owned the largest milking parlour in Europe, and was driven out of business by the collapse in prices and high interest rates in the 1990's. He now works as a herdsman for a local farmer, and helps teach primary school children about food production.

Graham Bigwood in the milking parlour, 2006.

Rounding up the milkers in the yard.

Jane Williams of Knights farm, Lydeard St Lawrence, spring lambing, 2007.

Jane Williams, Knights farm, 2007.

The lambing shed at Knights farm, 2007.

Derek and Tim Hobbs of Merlyns farm, near Nether Stowey, talking to the author about the future of farming. As County Council tenant farmers they are worried that their tenancy might not be renewed in four years' time, thus ending over fifty-five years of their family's farming at Merlyns.

Sheep arriving for sale at the old Taunton livestock market, spring 2007. With the new (out of town) market set to open at the end of 2007 another part of the tapestry of rural life, the weekly shopping trip to Taunton, will come to an end.

Jane Williams at Taunton market, 2007.

The auctioneer selling sheep, 2007.

Sheep auction at Taunton market, 2007.

A handful of horseshoe nails in the hands of travelling farrier, James Hazell, summer 2007.

James Hazell, farrier, summer 2007.

Richard Criddle, stock trader, London farm, Bagborough, 2006.

Emma Watson of BICAL Miscanthus, summer 2006. Although not universally welcomed around the hill, this new form of bio-fuel crop may prove part of the solution to the long term survival of the small family farm.

Phil Strickland and Julie Durston of the Fruit and Flower Basket farm shop, at Bishops Lydeard. One way to help small farmers is to make an effort to buy at least some of our weekly food shopping locally.

Henry Venn, general manager of agricultural machine sales at Ticknells in Cannington, the day after he was made redundant after thirty-four years with Ticknells, summer 2007.

Mark Peters and his family at their home at Westowe. Mark an ex-agricultural contractor now works as a Somerset paramedic. Like many others, Mark would have preferred to stay in farming, but with a family to support, it just didn't pay.

Mark Peters, ploughing part of his land with his much loved old Ferguson tractor.

The late Pete Thorne and his son Mick (Prickles) at Higher Aisholt farm, 2006.

Desmond Rockford, 83, ex-local farmer and a veteran of the Black Watch Commandos, on Remembrance Sunday, 2006.

Tim Russell and Andy Harris, Quantock Hills A.O.N.B. rangers.

Pulling a horsebox out of a ditch at Plainsfield, spring 2006.

'What is life if, full of care, we have no time to stand and stare'.

Hunt followers at Adscombe.

Ben Sherwin 41, Ballifants Farm Sale.

I met with Ben on the day of his farm sale. He had been informed that the owners of Ballifants wanted to take the farm back into their family's control in April of 2007. This left him to decide whether to continue farming for another winter, buying silage and other feed to keep his animals or to sell up. Reluctantly, like many farmers in his situation, he has decided to move on, and look for another place to farm.

If we were going to try and keep farming this winter we would have needed to buy 400 bales of silage, and to look for places for our stock, so all in all it just made more sense to sell up and look for a new place, and if I can't find another farm then I'll just have to get a job. It's very hard to make money from a small farm, the supermarkets control us because they are big, and we are all just individual producers. But, I have to be a businessman, if I can't make money then why do it? Some of these local lads are like sticks of rock, if you broke them open, they would have FARMER all the way through them. My father always said, farming is great, but you do need a proper job. Over sixty percent of road haulage is food, my message would be for people to try to buy as much as they can locally, or at least regionally, if people spent just another ten pounds a week on local food, British farming wouldn't need subsidies.

Ben Sherwin, at the Ballifants farm sale, Friday 29th September 2006.

Peter Huntley, the auctioneer preparing the sale paperwork at the Ballifants farm sale, September 2006.

Advertisements for future sales.

Enjoying some refreshments before the start of the sale.

Ballifants farm sale, September 2006.

A chance to catch up with news, before the start of the sale, Ballifants, 2006.

Prime Limousin cows, at the Ballifants farm sale, 2006.

Robert Norman of Northway farm, Halse, helping out at the Ballifants farm sale, September 2006.

Ballifants farm sale, September 2006.

Staking a claim to the dryer! Ballifants farm sale, September 2006.

Ballifants farm sale, 2006.

Ballifants farm sale, September 2006.

The auctioneer at work, with the benefit of a little technology, Ballifants farm sale, September 2006.

A well attended sale. Ballifants farm, September 2006.

Leaving the sale, Ballifants farm, September 2006.

Reed Combing at Waterhouse Farm

Richard White, Master Thatcher, whose family have been thatchers in Somerset since 1781, comes to Waterhouse farm each year to comb the dried wheat for use in reed thatching.

I bought this machine in 1978 for £2000. They don't make them any more, although there are a few about, the comber was made in Devon, but the thresher was made in Lincolnshire. I don't know how we would do this job without it; they haven't come up with anything modern to take its place. You can't have a machine that cuts the corn in the field and makes it into reed, because it just doesn't work. If you cut it in the field it would be too ripe for reed, you must cut it when the wheat is a bit green, and then stook it in the field and let it dry naturally for a week. Back in the 1970's the reed we were buying from the dealers was too ripe, the farmers were leaving it in the ground too long, that's why we bought the machine, so we could control the process more and keep up the quality of the reed we needed for the best thatching work. It's vital that it's cut at just the right time, it should be oily, you can feel it, even a blind man could tell good reed, by putting his hand in it. The crew are all getting on a bit, the oldest is Jack at 79, then George at 72, Dave is 66 and Doug is 64, I'm now 60, but we have got some of the youngsters coming through, up in the barn helping out today is my nephew who's learning the trade and he's just 18, and my son's got three sons who all want to learn the business so we should go on for a while yet.

Richard White, Master Thatcher, during the reed combing at Waterhouse farm, 2006.

Reed combing at Waterhouse farm, 2006.

Keeping an eye on the proceedings, 2006.

The reed comber, 2006.

One of the team, during a short break from feeding the machine, Waterhouse farm, 2006.

The air was thick with dust, the waste product of the reed combing at Waterhouse farm, 2006.

The machine sets a diabolical pace. Reed combing at Waterhouse farm, 2006.

Waterhouse farm, 2006.

Jack Rossiter, loading reed bundles into the barn at Waterhouse farm, 2006.

The reed comber feeding the small bale machine at Waterhouse farm, 2006.

After loading the lorry with a full load of reed bundles for thatching, Jim, Jack and Ingerma Rossiter, with lorry driver, Autumn 2006.

The Quantock Pony Fair

Richard Criddle, London Farm, Bagborough
A locally based stock trader, Richard was born and bred on the Quantock's. From his base at London farm, Richard has worked hard over the last twelve years to re-establish the annual sale of the Quantock ponies as a country fair, where people come to buy and sell, not just the ponies, but everything from tractors to chickens.

I'm 57 now so I've only got a few years to go, but you can't make farming pay any more, it makes me sad that I can't afford to have my son working with me here. He works on a farm up in Norfolk, and there's no way I could match what he earns up there, but he comes down and helps me when he can, he likes to help with the pony fair in October. I think the pony fair is helping to give people a sense of local identity, they come from miles around to see the ponies, last year we had over eight hundred people here, and this year we're hoping for even more. We'll be selling all sorts this year, lots of ponies, plus four hundred sheep and lots of farm equipment, it's a real old Quantock get together.

Driving sheep to the pony fair early morning, October 2006.

Kevin, farm labourer on the morning of the pony fair, October 2006.

Sorting sheep for sale at the pony fair, October 2006.

Checking the number of teeth, to tell the age of a ewe, at the pony fair, October 2006.

Signs of the times at the Quantock Pony Fair, October 2006.

Robert Venner, auctioneer, selling sheep at the pony fair, October 2006.

A mixed bunch! Sheep for sale at the pony fair, October 2006.

Queuing for the facilities, pony fair, October 2007.

Robert Venner, at work in the main field, pony fair, 2006.

Tractor for sale, pony fair, October 2007.

Old and young alike, all appreciate a well-built machine, pony fair, October 2007.

Sharing a laugh at the pony fair, October 2006.

Registering for the sale, October 2006.

Tom, 70 years old and still riding his hill ponies, October 2007.

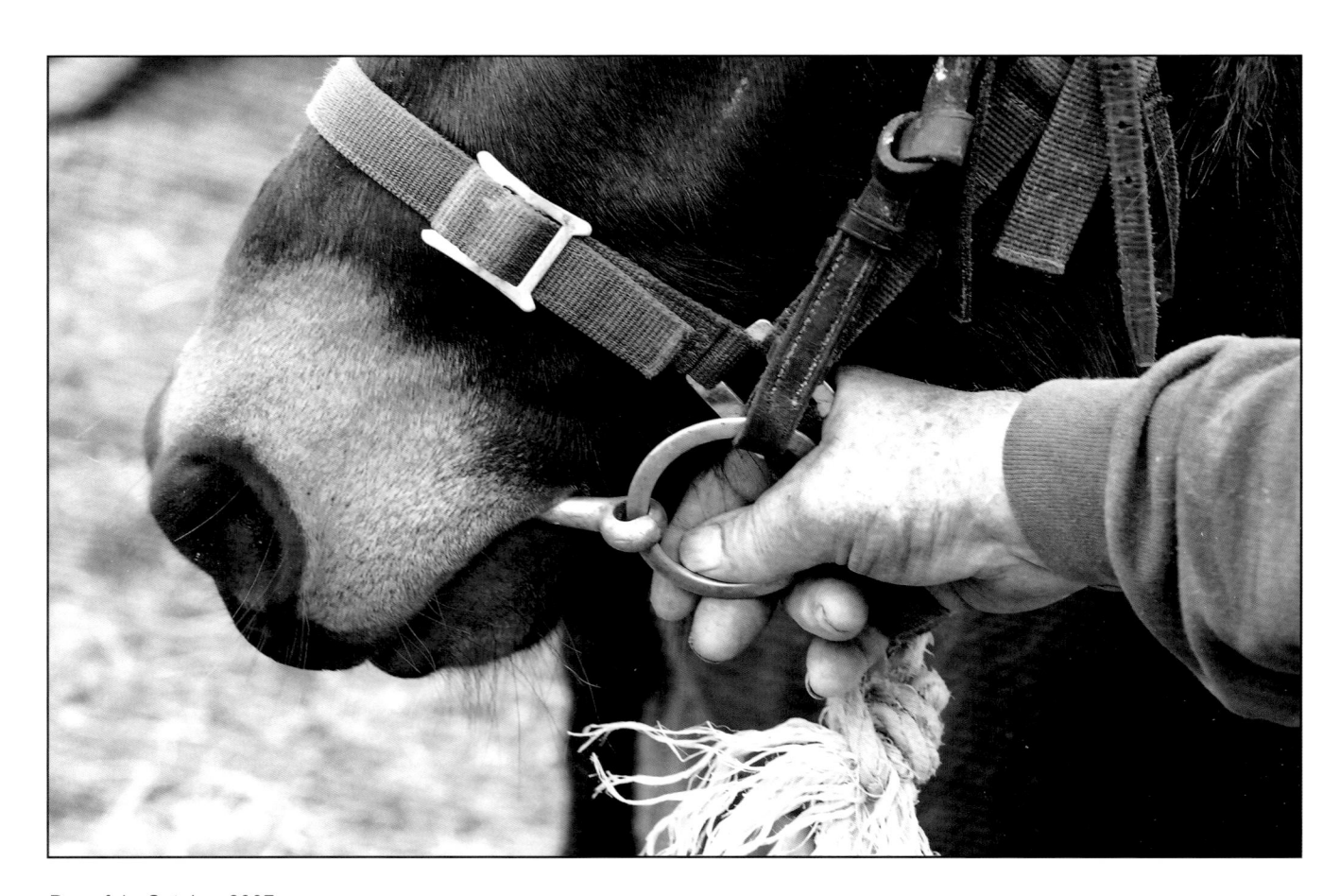

Pony fair, October 2007.

Robert Venner, auctioneer, selling the hill ponies at the Quantock Pony Fair, October 2006.

*'Trackway camp and city lost
salt marsh where now is corn
old wars, old peace
old arts that cease
and so was England born'*
Kipling

Jack Rossiter, harvesting wheat with vintage cutter/binder, July 2006.

Bluebell woods on the Quantocks.

The Quantock beeches, Winter 2006.

Pyleigh House farm, near Lydeard St Lawrence, spring 2006.

Ingerma Rossiter, stacking wheat 'stooks' at Waterhouse farm, July 2006.

The old 'pole' barn at Merlyns farm, near Nether Stowey, 2006.

Mile marker outside Cothelstone Manor.

Vintage tractor driving vintage reed comber, at Waterhouse farm, 2006. Still a commercial activity. After fifty years of use, the machine is still going strong and working as well as when it was first built.

'Cultivators of the earth are the most valuable of citizens. They are the most vigorous, the most independent, the most virtuous and they are tied to their country and wedded to its liberty and interests by the most lasting bands.'

Thomas Jefferson

'I am convinced that it is better for a writer to know a little bit of the world remarkably well, than to know a great part of the world remarkably little'.

Thomas Hardy